EAT
TO **PREVENT**
AND **CONTROL**
DIABETES

How Superfoods Can Help You Live Disease Free

(Extract Edition)

LA FONCEUR

Eb
emerald books

Eb

emerald books

Copyright © La Fonceur 2021
All Rights Reserved.

This book has been published with all efforts taken to make the material error-free. The information on this book is not intended or implied to be a substitute for diagnosis, prognosis, treatment, prescription, and/or dietary advice from a licensed health professional. Author doesn't assume and hereby disclaim any liability to any party for any loss, damage, or disruption caused by errors or omissions, whether such errors or omissions result from negligence, accident, or any other cause.

While every effort has been made to avoid any mistake or omission, this publication is being sold on the condition and understanding that neither the author nor the publishers or printers would be liable in any manner to any person by reason of any mistake or omission in this publication or for any action taken or omitted to be taken or advice rendered or accepted on the basis of this work.

Dear reader,

The aim of **Eat to Prevent and Control Diabetes** is to help reduce your dependence on medicines by providing you with in-depth knowledge of common chronic diseases as well as the best food options that prevent and control diabetes naturally.

Eat healthily, live happily!

Master of Pharmacy, RPh

and Research scientist

CONTENTS

Introduction	7

UNIT 1 EAT TO PREVENT DISEASE — **9**
Role of Food Therapy in Preventing and Controlling Disease — 10

UNIT 2 DIABETES: PREVENTION AND CONTROL — **23**
Everything You Need to Know About Diabetes — 25
10 Foods That Increase Your Diabetes Risk — 40
10 Best Foods to Prevent And Control Diabetes — 50

UNIT 3 DIET PLAN — **63**
Diet Plan to Control Diabetes — 65
Diet Plan to Control Diabetes + Hypertension — 68
Diet Plan to Control Diabetes + Arthritis — 70

UNIT 4 RECIPES — **73**
Lunch — 74
  Chole Masala — *75*
  Non-Fried Oats Bhature — *79*

Note From La Fonceur — 81

References — 82

Important Terminology — 84

Abbreviations — 86

About The Author — 87

All Books By La Fonceur — 88

Connect With La Fonceur — 89

INTRODUCTION

Nowadays, diabetes has become quite common. One in every family has diabetes. People have started considering diabetes as part of life, which is not good. The lifestyle we are leading today - high intake of processed foods, frequent eating out, smoking, and alcohol, there is a 70% chance that you will have either high blood sugar levels or high blood pressure or both by your 50s.

A disease state in the body means your immune system is constantly busy fighting the disease, soon your immune system loses its effectiveness and becomes weak. If another disease strikes, your immune system is unable to fight, this can have life-threatening consequences. It is very important to start as early as in your 20s to take care of your health. Make your body strong enough to fight any disease naturally.

More diseases mean more medicines. Being from a pharmacy background, I can assure you that dependence on medicines is not good. Medicines prescribed in disease have side effects. To reduce side effects, you are often prescribed with another set of medicines that treat the side effects of your primary medications, but they also have side effects, for which again some other medications are required, so basically, this cycle continues. But there is a solution! You can include foods in your diet that have the same effect as your medications. By regular intake of

INTRODUCTION

these foods, you can heal your body and increase your immunity to fight disease naturally.

The objective should be to prevent disease, and preparation starts in your 20s. What you eat in your 20s affects your 50s. To prevent a disease, you must have a thorough knowledge of the disease, such as why it happens? How does it affect your body? What exactly does happen in your body in the event of a disease? What are other health problems that can be caused by a particular disease?

In *Eat to Prevent and Control Diabetes*, all these topics will be discussed in detail. You will learn about foods that boost your immunity, superfoods that can protect you from diseases, foods that reduce inflammation in your body, and food combinations that you should eat for maximum health benefits.

You will also learn everything about diabetes. To prevent this disease, which foods and lifestyle options should you avoid, and which ones should you adopt? What should be your strategy to prevent and control diabetes. What are the foods that mimic your medication's mechanism of action and can help lower your blood sugar levels? What are the key points that you can follow to prevent diabetes and how to get rid of it?

You will also discover some healthy and tasty recipes that have all the healthy ingredients and still very tasty. These recipes will strengthen your immunity as well as satisfy your taste bud. Get ready for a healthy tomorrow.

UNIT 1

EAT TO PREVENT DISEASE

1
ROLE OF FOOD THERAPY IN PREVENTING AND CONTROLLING DISEASE

Role of food therapy in preventing and controlling disease

If you have your blood sugar levels checked for the first time, and your report says that your blood sugar is high, your doctor will not first prescribe you the medicines. Instead, your doctor will give you a three-month time so that you can control your sugar with diet and lifestyle modifications. If the blood sugar is still not controlled, then only you will be prescribed with medicines to control high sugar levels.

You know why? Because medicines treat disease, but they can cause strong side effects. The stronger the drug, the stronger will be its side effects. It doesn't mean you should stop taking medicines without informing your doctor. Never stop your medication without consulting your doctor because some medicines have withdrawal effects, which can even worsen your disease condition if you stop taking them suddenly.

So, what is the solution? The solution lies in good management. You can prevent or manage the disease only when you have comprehensive knowledge about that disease. Everything is on your hand, and you are the commander of your life and your disease condition. With the correct nutrition and healthy lifestyle, you require fewer medicines, shorter therapy duration, and minimum side effects.

When it comes to disease management, there are lots of misconceptions associated with it. Let's first clear these misconceptions:

#1 Misconception

I am young, and I don't have any disease, I have plenty of time to live without worrying. I will worry about diseases

EAT TO PREVENT AND CONTROL DIABETES

when I will hit my 50. Till then, my motto is *You Only Live Once*.

Actually, you only die once but live every day, so make your every day disease-free. The age of 20 to 40 is your key to make your 50+ years healthy and happy. The way you treat your body between these years, its effect is seen in your old age. These are your sowing years, eat as many healthy foods as you can during these years, and reap the benefits in your 50+ years. Strictly avoid smoking, alcohol, and other drugs that deteriorate your health internally. The harm never visible during your 20s and 30s, but it has life-threatening consequences soon after you hit your 50s or nowadays even in your 40s. Eat junk foods but only to satisfy your taste bud, definitely not to fill your tummy.

#2 Misconception

I am a very health-conscious person, and I believe nature has all the solutions. Although I have been diagnosed with a disease, but I believe I don't need medication. I can heal myself naturally with healthy food and a good lifestyle.

If you have been diagnosed with a disease means the harm has unknowingly already been done. Keep in mind medicines are not the enemy, just they are not a natural food. Sole dependence on medicines is not good; at the same time, completely abandoning medicines when your body needs them is also not right. No doubt, healthy foods, and healthy lifestyle choices can heal you much faster, but you definitely need medication to treat a disease. With healthy foods, you can heal yourself faster, and your body recovers more quickly, so you need a shorter course of therapy that simply means lesser side effects.

#3 Misconception

Last time when I had these symptoms, my doctor prescribed these medicines. Now again, I feel the same problem, I should take the same medicines as the doctor had prescribed me last time.

Avoid self-medication. In a condition of reoccurrence of any disease, medicines may be the same, but with different doses. Don't be a doctor yourself. Self-medication may result in an overdose, which can lead to toxicity and other life-threatening consequences. Seek advice from your doctor every time you are not feeling well and ask him/her directly if it is safe to take the same medication again when the symptoms occur. Always ask your doctor about what should be the diet in managing your disease? Ask your pharmacist if there is any food that you should avoid while taking the prescribed medicine.

#4 Misconception

I was on medication, and my condition has improved. Although my doctor had prescribed me a 3-month course of medicines, I was feeling fine, so after two months, I stopped taking the medicines.

This is never advised. Maybe with your healthy diet and lifestyle, you have recovered faster than others, but you should never leave your medication course in between

EAT TO PREVENT AND CONTROL DIABETES

without consulting your doctor. Even if your symptoms are relieved with initial medications, you need the full course to treat the disease completely. Otherwise, it will reoccur, and as it has not been treated in the initial stage, it will reoccur with more severeness. Abruptly discontinuing some medicines produce withdrawal effects in the body and worsen the disease condition. Instead of quitting your medicines, you should inform your doctor about your improvement. Your doctor will gradually reduce the dose of the same medication and complete the course sooner than before, or he will advise you to complete the full course depending upon your disease type and your condition.

#5 Misconception

I am taking medicine for my illness, and medicine is doing its job. It will cure me, I don't have to worry too much about nutrition and all.

Food and lifestyle play a huge role in managing any disease. If your diet is not healthy and your lifestyle is also not good, then your condition may worsen despite regular medication. Foods that boost immunity prepare your body to fight the disease and heal your body. A healthy lifestyle removes the burden from your body; hence your body can completely focus on treating the disease.

#6 Misconception

Some diseases like diabetes, high blood pressure, arthritis, etc. come naturally with age. You cannot escape from these diseases. Every other person I know

has one of these diseases, so this is pretty normal at my age.

It may be common but indeed not normal. This is the biggest myth that some diseases naturally come with age. With age, our body becomes a little weaker, but most of these diseases are the result of our poor diet and poor lifestyle. It's time to stop letting these diseases become part of your life and build your body with healthy foods and a healthy lifestyle in such a way that these diseases never even touch you, and if you already have these diseases, then they can be controlled.

DISEASE MANAGEMENT

What is a disease?

A disease is a condition of disturbances in the normal structure or function of your body. When something goes wrong with the normal function of your body, your body gives signals in the form of signs and symptoms that something is going wrong inside the body. This is where your responsibility begins. With proper medication, healthy foods, and healthy lifestyle, you can prevent and treat disease.

Disease Is mainly managed with medications and foods and lifestyle modifications. Let's understand the role of each.

ROLE OF MEDICATIONS

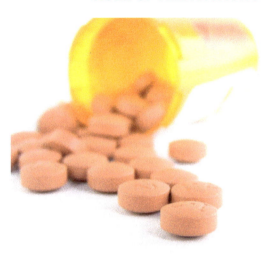

Prescribed medications play a vital role in the treatment. Generally, medicines work in three ways:

1. To reduce symptoms like pain, nausea, and fever.

2. To treat the disease.

3. To reduce or treat the side effects arising due to the use of medicines in curing the disease. Such as antacids are usually prescribed with high dose medications as these medications can cause acidity in the body.

It is important to take the medicines at the same time every day. Medicines take their time (onset time) to show their effect on the body. Taking medicines at the same time every day ensures that its active ingredient will be available in the body uniformly throughout the treatment.

ROLE OF LIFESTYLE CHOICES

Poor lifestyle choices give a burden to your body. In simple terms, these are the inducer of many diseases. Unhealthy lifestyle choices weaken your body, lower your immunity, and make you susceptible to many diseases.

Example of unhealthy lifestyle choices:

Stress

Smoking

Alcohol

Unhygienic habits

Inadequate sleep

When it comes to poor lifestyle choices, you may have found plenty of discussions about smoking, alcohol, and inadequate sleep, but we often take other poor lifestyle choices such as stress and unhygienic habits lightly.

EAT TO PREVENT AND CONTROL DIABETES

Stress is a key contributor to many diseases. When you are under stress, your body releases the stress hormone cortisol, which causes your heart to pump faster and raises your blood pressure. After the stressful time has passed, your body releases lower amounts of cortisol. Your heart and blood pressure return to normal. But if you are under constant stress, the consistently high levels of cortisol in your body can cause many health problems. Take out at least 2 hours for yourself every day, do nothing during this time, just relax. Just two stress-free hours in a day gives your body enough time to normalize all its functions and systems.

Unhygienic habits such as not washing your hands before eating and after using the washroom, and touching an open wound allow germs to enter the body. As a result, your immune system keeps busy fighting these germs and with time immune system weakens. When your immune system becomes weak, it can't protect you from major and serious diseases. So, don't stress out your immune system. Already there is so much pollution in the environment with which your immune system fights daily, so do not give it more burden. Maintain good hygiene and keep your immune system healthy. Whenever you come from outside, first wash your hands with soap water. This habit will protect you from many diseases. Also, 5-Second rule is a big myth. Even a brief exposure of the floor can contaminate your food with E. coli, salmonella, and other bacteria in under five seconds.

What should be your focus points to prevent disease?

If you keep your immune system and digestive system healthy, you greatly lower your risk of diseases.

Avoid Immune weakening lifestyle choices

- Stress
- Smoking
- Drinking Alcohol
- Consuming narcotics such as Cannabis

Adapt the immune-boosting lifestyle choices

- 7-8 hours of sleep
- Washing your hands frequently
- Expose yourself to early morning sunlight
- Yoga
- Taking a walk after lunch and dinner

Avoid immune weakening foods

- Trans fats
- Processed foods
- Canned foods
- Refined carbohydrates
- Foods high in sugar

EAT TO PREVENT AND CONTROL DIABETES

Add immune-boosting foods in your diet

- Foods rich in vitamin C
- Foods high in zinc
- Foods that have anti-inflammatory effects

ROLE OF FOODS

Foods play a massive role in every stage of disease management. These roles include:

1. To prevent the disease.

2. To shorten the therapy period.

3. To control the disease.

4. To prevent the reoccurrence of the disease.

Body is nature's product, and your body loves natural things like food. Plant-based healthy foods can prevent various diseases and autoimmune disorders and help build your body strong enough to fight off any disease. Plant-based healthy foods heal your body, reduce your dependency on medicines, and add disease-free years to your life.

Why food therapy is the best therapy?

Medicines work on the disturbance that has been arisen in your body dues to disease, while foods work on the root cause and strengthen your body to fight with the disease naturally. Moreover, foods have no side effects. A simple rule is to include vegetables and fruits of every color in your diet; it will protect you from countless diseases.

We are so concerned about avoiding unhealthy foods that it has become quite stressful now. The more you try to run away from them, the more you crave. If you do not eat unhealthy foods, but your diet lacks essential nutrients, then there is no health benefit of avoiding unhealthy foods. To stay healthy, incorporating healthy foods into your diet is more important than just avoiding unhealthy foods. It is time to focus on what you should eat, not what you should not eat. Do not modify your diet for weight loss; instead, balance your diet for a healthy and disease-free life.

UNIT 2

DIABETES: PREVENTION AND CONTROL

1
EVERYTHING YOU NEED TO KNOW ABOUT DIABETES

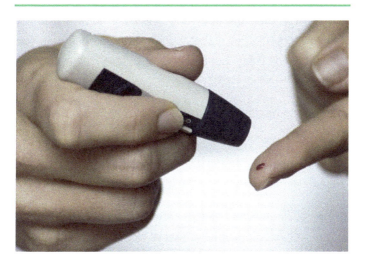

Everything you need to know about diabetes

Diabetes is probably the most common chronic disease; 1 in 11 of the world's adult population is living with diabetes. It's like a slow poison. It slowly affects the other part of your body and a major cause of kidney failure, blindness, and heart attacks. The worrying part is people don't take diabetes seriously. The younger generation is not well educated about it, and affected people are highly dependent upon medicines, giving less effort on diet and exercise front. Diabetes doesn't come with age, but it comes with a bad lifestyle and diet deficient in healthy

EAT TO PREVENT AND CONTROL DIABETES

nutrients. If you have type 2 diabetes, you don't need to live with it; diabetes is reversible. Early diagnosis, a healthy diet, physical activity, and medications can help you reverse diabetes. If you have diabetes for a long time and you are on high dose medications, your aim should not be just to avoid sugary foods. You should aim to eat foods that mimic the action of your anti-diabetic drugs and exhibit similar effects in the body and also have no side effects. Regular screening of diabetes complications can help you prevent and treat complications before they become severe.

Here are some dangerous facts about diabetes that no one talks about:

- According to the International diabetes federation, diabetes caused 4.2 million deaths in 2019.

- 374 million people are at increased risk of developing type 2 diabetes, according to IDF.

- According to WHO, diabetes is a major cause of kidney failure, heart attacks, stroke, and blindness.

- Diabetes was the seventh leading cause of death in 2016, according to WHO.

Don't just accept diabetes as a new normal; it may be common but not normal. Let's prevent and control diabetes through natural ways. But for that, first, you need to have thorough knowledge about every aspect of diabetes so that with a little help from health professionals (who can guide you at every step where you will have confusion), you can prevent and control diabetes without medicines.

What is diabetes?

Diabetes is a chronic disease, it occurs either when the body does not produce insulin or does not effectively use the insulin it produces. Insulin is a hormone that helps the body use sugar for energy. Hyperglycemia is the medical term for high blood sugar (glucose) levels. It is a common effect of uncontrolled diabetes. Over time, hyperglycemia leads to severe damage to many of the body's systems, especially the nerves and blood vessels.

Types of diabetes:

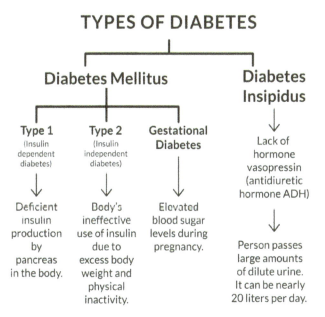

DIABETES INSIPIDUS

People with diabetes insipidus have normal blood glucose levels. The only similarity between diabetes mellitus and diabetes insipidus is excessive urination. Diabetes insipidus occurs when kidneys cannot balance fluid in the body. It happens due to some damage in the pituitary gland, which releases the anti-diuretic hormone (ADH), also known as vasopressin. ADH or vasopressin enables the kidneys to retain water in the body. In the absence of vasopressin, kidneys excrete too much water. This causes frequent and excessive urination and can lead to dehydration.

DIABETES MELLITUS

Diabetes mellitus, or simply diabetes, is a disorder that is characterized by abnormally high blood sugar (glucose) levels because the body does not have enough insulin to meet its needs. It is caused either because the body doesn't produce enough insulin or because the body is ineffective in using insulin. In diabetes, urination and thirst are increased and damages the nerves and tiny blood vessels that cause health complications, especially in the kidneys and eyes.

What happens in your body in diabetes?

When you eat food containing carbohydrates, your body breaks them down into sugar (glucose) and send that to your bloodstream. The rise in glucose triggers the pancreas to release insulin into the bloodstream. Insulin signals muscle, liver, and fat cells to take in glucose from the blood. These cells then convert glucose into energy or store it for later use. In diabetes mellitus, your body doesn't use insulin as it should, which results in too much glucose in your blood.

Type 1 diabetes

(Insulin-dependent diabetes)

Type 1 diabetes is an autoimmune condition in which the pancreas cannot produce insulin. Typically, the immune system protects your body against the attack of bacteria and viruses by fighting against them. An autoimmune condition is a condition in which your immune system mistakenly attacks your body's cells. In type 1 diabetes, the immune system destroys the beta cells of the pancreas that produce insulin, making the pancreas unable to produce insulin.

Insulin is a hormone that regulates the blood sugar levels in your body. In the absence of insulin, your body can't use or store glucose for energy. The glucose stays in your blood, and your blood sugar or blood glucose levels become too high (hyperglycemia). Persistent high glucose levels result in diabetes and can lead to complications affecting your kidneys, nerves, eyes, and heart. A person who has type 1 diabetes requires daily insulin injections to control blood glucose.

The condition usually appears in children and young people, so it used to be called juvenile diabetes.

Type 2 diabetes

(Insulin-independent diabetes)

Type 2 diabetes is the most common form of diabetes that primarily occurs due to obesity and lack of exercise. It is characterized by high blood sugar (hyperglycemia) and insulin resistance.

Insulin resistance means your pancreas is releasing insulin as it should, but the cells in your muscles, fat, and liver start resisting the signal given by insulin to take glucose out of the bloodstream for making energy. This results in too much glucose in your blood, known as prediabetes.

In a person with prediabetes, the pancreas works increasingly hard to release enough insulin to overcome the body's resistance and to keep blood sugar levels down. Over time, the pancreas' ability to release insulin begins to decrease, which leads to the development of type 2 diabetes.

Reason for Insulin resistance

The driving forces behind insulin resistance are excess body weight, too much fat in the abdominal area, and a sedentary lifestyle, while genetics and aging also play a role in developing insulin resistance.

GESTATIONAL DIABETES

Gestational diabetes is a condition in which your blood glucose levels become high during pregnancy. Body goes through different changes during pregnancy, such as weight gain and changes in hormones, which affects the body's cells' ability to respond to insulin effectively. Most of the time pancreas can produce enough insulin to overcome insulin resistance, but some pregnant women cannot produce enough insulin and develop gestational diabetes. Mostly, gestational diabetes goes away soon after delivery. Women who develop gestational diabetes during pregnancy are at higher risk of developing type 2 diabetes later in life.

Here is everything about diabetes mellitus that you need to know to prevent and control it:

SYMPTOMS OF DIABETES

Excessive urination (polyuria): To get rid of excess glucose, kidney makes more urine than usual. Daily urine output can be more than 3 liters a day compared to the normal urine output about 1 to 2 liters.

Excessive thirst (polydipsia): Too much glucose forces kidneys to work overtime. Kidneys pull water from tissues to make more urine to help pass the extra glucose from your body, which makes you dehydrated. This usually makes you feel very thirsty.

Fatigue: Because the body's cells do not get enough glucose to make energy.

EAT TO PREVENT AND CONTROL DIABETES

Weight loss (in Type I diabetes): Because of dehydration caused by excessive urination and loss of calories from the sugar that couldn't be used as energy.

Constant hunger: Because the body can't convert the food you eat into energy.

Blurred vision: High blood sugar causes body water to be pulled into the lens inside the eye, causing it to swell.

WHY IS DIABETES DANGEROUS?

Uncontrolled diabetes can lead to potential health complications, including:

Retinopathy (Eyes damage): High blood sugar levels can weaken and damage the small blood vessels of the retina, which can cause visual disturbance and can even lead to blindness.

Neuropathy (Nerve damage): Constant high blood sugar can damage nerves that typically results in numbness, weakness, tingling, and burning or pain, usually in the hands and feet (diabetic foot).

Nephropathy (kidney damage): Over time, diabetes can damage the small blood vessels in the kidneys, which can lead to kidney failure, and the person may require dialysis or a kidney transplant.

Ketoacidosis (mostly in Type 1 diabetes): When there isn't enough insulin in the body to convert glucose into energy, your body starts breaking down fat for energy. This process produces a build-up of acidic substances called ketones to dangerous levels in the body, eventually leading to ketoacidosis.

Heart disease: Over time, the high blood sugar levels can damage the blood vessels that maintain the heart function, causing them to become stiff and hard. A high-fat diet can cause a build-up of fats and cholesterol on the inside of these blood vessels, which can restrict blood flow. This condition is known as atherosclerosis. Atherosclerosis conditions can reduce blood flow to the heart muscles (which causes angina) and brain (which causes stroke) or can damage the heart muscle, which can result in a heart attack.

DIABETES: PREVENTION AND CONTROL

Diabetes condition can be effectively managed by

Diet

Medication

Exercise

Before going further, let's clear some terms associated with diabetes:

Glycemic index

You must have heard about low glycemic foods and high glycemic foods, but what is exactly the glycemic index?

The Glycemic index helps you differentiate between good carbohydrates and bad carbohydrates for diabetes. Not all carbohydrates are the same. Type of carbohydrates like complex carbohydrates takes longer to break down into glucose and slowly absorbed and metabolized that cause a slower rise in blood sugar. This type of carbohydrates doesn't give you a sudden spike of sugar levels and considered as good carbohydrates. These are categorized as low-glycemic foods; they help maintain good glucose control. Foods that have the glycemic value 55 or less are good for diabetes—for example, whole grains and beans.

Simple carbohydrates such as sugar and highly processed and refined carbohydrates such as pastries and cakes are considered as high glycemic foods. They speedily break down into glucose and quickly absorbed, causing a rapid rise in blood sugar. Repeated spikes in blood sugar lead to an increased risk for type 2 diabetes.

Hypoglycemia

Hypoglycemia condition is often caused by diabetes treatment. Hypoglycemia is the opposite of hyperglycemia. It is a condition in which your blood sugar levels are lower than normal. Certain diabetes medicines or too much insulin may cause your blood sugar levels to drop too low. It is a reversible condition and can be treated by consuming high-sugar foods such as fruit juice or honey. If you are on diabetes medication, you should pay attention to hypoglycemia symptoms, which include confusion, shakiness, and dizziness. If left untreated, hypoglycemia can get worse and can even cause seizures, coma, and death. Always keep glucose tablets with you in case you experience hypoglycemia.

ROLE OF MEDICATIONS IN DIABETES

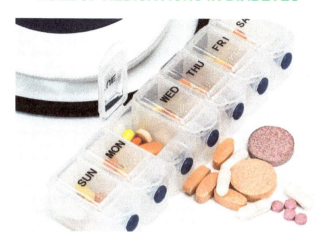

How do diabetes medicines work?

Your most common and first choice of drug in diabetes (metformin) doesn't stimulate insulin secretion by beta cells in the pancreas; instead, it enhances the ability of your tissues to take glucose out of the bloodstream and convert it into energy, especially in muscles. Additionally, it lowers the production of glucose by the liver. It is a drug of choice because it doesn't cause weight gain and hypoglycemia. The common side effect of metformin is diarrhea. Don't just stop taking your medicine because of diarrhea; instead, eat yogurt, beans, an apple, or a banana (not more than one banana in a day) to prevent diarrhea. Make sure to drink plenty of water to prevent dehydration caused by diarrhea.

Another class of drugs (sulfonylureas), reduces high blood sugar by increasing insulin secretion by beta cells in the pancreas. Additionally, it increases cells' sensitivity to insulin, which increases the efficiency of the body's cells to take out glucose from the bloodstream. It also increases the availability of insulin in blood by reducing the degradation of insulin in the liver. The side effects of this class of drugs are weight gain and hypoglycemia. It is very important to keep an eye on your hypoglycemia symptoms, which include shakiness, sweating, dizziness, confusion, irritability, and loss of consciousness. Severe hypoglycemia can potentially lead to coma. Take glucose tablets (total 15g or ask your doctor for exact amount) or foods high in glucose like one tablespoon of honey or sugar or 3-4 raisins to treat hypoglycemia immediately. Ask your doctor to adjust the dose of your medicines if you experience hypoglycemia.

ROLE OF EXERCISE IN DIABETES

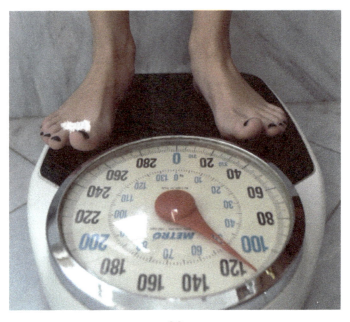

Obesity and diabetes connection

Obesity is a common reason for type 2 diabetes. Just by losing weight, you can prevent the onset of diabetes. If you are in the prediabetes stage, you can even reverse diabetes by losing weight, and by including hypoglycemic and weight-loss-friendly foods in your diet.

Excess fat may contribute to insulin resistance. It is because when your fat cells that store extra fat become too large, they stop storing fat. Additional fat starts storing in muscles, liver, and pancreas, making these organs resistant to insulin, and they stop responding to the signal given by insulin to take glucose. Moreover, fat cells decrease the secretion of adiponectin, a protein hormone that helps in the breakdown of fat. In simple terms, adiponectin is your fat-burning hormone. High adiponectin levels can protect you against insulin resistance, diabetes, and heart disease. The more you lose weight, the higher your adiponectin levels.

How can weight loss prevent the onset of diabetes?

In type 2 diabetes, insulin production decreases over a sustained period, and the process is rather slow in comparison to type 1 diabetes. It is possible that a strict diet and exercise regime, leading to weight loss, and may delay or even prevent the onset of diabetes. The key is to diagnose the diabetes condition before beta cells function deteriorates.

After the age of 40, you should get your sugar levels tested every year for early diagnosis of high blood sugar levels. If you are overweight, losing weight is your first and most crucial step. It will not only save you from diabetes but can protect you against many diseases. You don't need to lose weight in a short time. Start eating diabetes-friendly foods and start with moderate exercise, soon your body will get habitual of your new diet and exercise regime. It should not be your short-term goal but a new lifestyle.

Ideal weight target to prevent insulin resistance

Body mass index: 25 kg/m^2

Waist circumference: Less than 100 cm

ROLE OF DIET IN DIABETES

A healthy diet plays an important role in preventing and managing diabetes. Preventing diabetes isn't just about avoiding foods that can spike your blood sugar levels, it's also about choosing the right foods that naturally prevent diabetes. In diabetes, moderation and frequency are the keys, you can still eat your favorite foods, but you might need to eat them less often or eat smaller portions.

To prevent and control diabetes:

- Avoid sugary foods that directly raise your blood sugar levels.

- Avoid refined carbohydrates that quickly break down into glucose sugar and raise your blood sugar levels.

- Avoid foods that increase your risk of insulin resistance.

- Avoid foods that increase cholesterol in the body.

- Avoid lifestyle choices that increase the risk of developing diabetes.

- Avoid foods that increase your risk of developing diabetes complications.

Now that you know what type of foods you need to avoid, let's see in the next chapter, which are the top 10 foods and lifestyle choices that you should avoid to prevent and control diabetes.

2

10 FOODS THAT INCREASE YOUR DIABETES RISK

10 foods that increase your diabetes risk

Below are the 10 foods that can increase your diabetes risk:

1. Saturated fats

Saturated fats increase the risk of developing type 2 diabetes more than unsaturated fats. Foods that are high in saturated fats, such as butter, cheese, cream, and processed foods like cakes and biscuits can cause high levels of LDL-cholesterol, which increases the risk of cardiovascular diseases in people with diabetes. LDL-cholesterol transports from the liver to the cells, and fat build-up inside muscle cells lowers the insulin response to glucose and increases blood sugar levels, increasing the risk of developing diabetes. Foods like butter, coconut oil, palm oil, and cheese have high amounts of saturated fat.

2. Starchy foods

Starchy foods such as white rice, boiled potatoes, and pasta are high in glycemic index. These foods are quickly digested and absorbed, causing a rapid rise in blood sugar levels. The best way to reduce their effect is to replace them with healthier alternatives. These healthy alternatives contain fiber that is low in the glycemic index. You can replace white rice with brown rice, boiled white potatoes with sweet potatoes, and white pasta with whole wheat pasta or durum wheat pasta. Even though these healthy alternatives provide fiber, you should eat them in moderation.

3. Packaged fruit juices

Packaged fruit juices are high in fructose and low in fiber. It gives a sudden spike in blood sugar and is less nutritious than freshly squeezed juices. Unfortunately, even the healthiest packaged fruit juice in the market can increase insulin resistance and increase your risk of developing diabetes. In fact, not only packaged fruit juices, even fresh fruit juices are not healthy in comparison to fresh fruits. Most of the fibers, vitamins, and anti-oxidants are removed while filtering. So, it is best to eat whole fresh fruit instead of fruit juice.

4. Hydrogenated oils

Hydrogenated oils are mainly present in packaged food items such as peanut butter, french fries, margarine, ready-to-use dough, and readymade baked foods. Hydrogenated oils are nothing but healthy vegetable oils that are converted into unhealthier form by the food industry. Vegetable oils are liquids at room temperature, food manufacturers chemically alter the structure of vegetable oils and turn them into solid or spreadable form by adding hydrogen in them. As a result, trans fats are formed. Trans fats increase insulin resistance by affecting cell membrane functions and increase your risk of developing type 2 diabetes. They are a major contributor to heart disease because they are highly inflammatory and can increase your LDL-cholesterol (bad) while lowering your HDL-cholesterol (good).

EAT TO PREVENT AND CONTROL DIABETES

5. Tobacco use

Tobacco use may fluctuate your glucose levels by altering the way your body uses glucose. Tobacco contains an addictive chemical called nicotine that increases insulin resistance, which can lead to type 2 diabetes. Tobacco stimulates the secretion of a steroid hormone called cortisol, which increases the production of glucose by the liver and makes fat and muscle cells resistant to the action of insulin. The more you smoke, the higher your risk of developing diabetes. Smokers have almost double the risk of developing diabetes compared with people who don't smoke.

6. Alcohol

Excess alcohol intake increases your risk of developing type 2 diabetes. Alcohol contains a lot of calories, which can make you obese. Obesity increases insulin resistance, which can lead to diabetes or can worsen your diabetes condition. Another disadvantage of drinking alcohol is that it produces a synergic effect and causes hypoglycemia by interacting with some of your prescribed anti-diabetic medicines such as sulfonylureas. Typically, the liver releases stored glucose when blood sugar levels drop, to maintain normal blood glucose, and to prevent hypoglycemia. But when you drink alcohol, it interferes with the way liver works and reduces the liver's ability to recover the dropped blood glucose levels. It results in hypoglycemia.

7. Soda

Soda and sugar-sweetened energy drinks can increase your diabetes risk, and if you already have diabetes, you must totally avoid them. The high sugar content of these drinks causes rapid spikes in blood sugar levels. These sugary drinks contain lots of calories, which can make you obese. The excess body weight makes your muscles, liver, and fat cells resistant to the insulin's signals of grabbing glucose out of the bloodstream. The high sugar levels in your blood make your pancreas to release more and more insulin to overcome the body's resistance and to maintain the blood sugar levels normal. Over time, this affects the pancreas's ability to make sufficient insulin, and your blood sugar begins to rise, and you develop diabetes.

8. Full fat dairy

Full-fat milk and milk products can increase the levels of cholesterol in the blood and lead to a higher risk of cardiovascular diseases. High-fat content can also lead to insulin resistance. Avoid eating full-fat milk, butter, full-fat yogurt, full-fat ice cream, and cheese. Even skimmed milk contains carbohydrates and can affect your blood sugar levels, but you should not completely avoid milk because it contains nutrients, which are a must for your body to function properly. It is better to remove other high calories and sugary food sources from your diet than milk.

9. Salt

It is crucial to maintain normal blood pressure in diabetes. Salt does not directly affect blood glucose levels, but you should limit your salt consumption for managing diabetes efficiently. Too much salt can raise your blood pressure. High blood pressure with diabetes increases your risk of cardiovascular diseases. You should maintain your blood pressure less than 130/80 mm Hg. You must limit your table salt consumption to 5g or one teaspoon per day to prevent and control diabetes.

10. Certain drugs

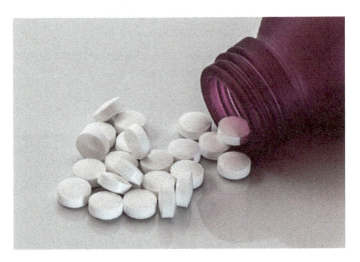

Certain drugs such as corticosteroids and pain-relieving non-steroidal anti-inflammatory drugs (NSAIDs) are contradicted in diabetes. Oral Corticosteroids can increase blood glucose levels and cause insulin resistance by reducing the sensitivity of the cells toward insulin. Corticosteroids can worsen diabetes conditions; this is why people with diabetes, as well as individuals with pre-diabetes, should avoid them.

People with diabetes who are receiving sulfonylureas drugs should avoid taking a high dose of pain-relieving non-steroidal anti-inflammatory drugs (NSAIDs) such as ibuprofen. One of the side effects of sulfonylureas is hypoglycemia means it lowers the blood sugar levels than the normal range. NSAIDs affect the ion channel functions of beta cells that secrete insulin. When you take NSAIDs together with sulphonylureas, it induces hypoglycemia.

CONCLUSION

These were the foods and lifestyle choices that you must avoid or at least limit the consumption to prevent diabetes, but as I said before, just avoiding harmful foods is not enough in managing diabetes. You must eat the right nutrition; in fact, eating foods that naturally prevent and even treat your diabetes is more important than just avoiding harmful foods. Diabetes-friendly foods not only can help control blood sugar levels, but some of them can even repair beta cells and can increase your insulin sensitivity. With regular consumption of these foods, your body naturally builds a defense system against diabetes, and you control diabetes without medicines or with a reduced dose of your medications.

Now let's see which are the top 10 best foods that can help you prevent and control diabetes without medicines.

3

10 BEST FOODS TO PREVENT AND CONTROL DIABETES

10 best foods to prevent and control diabetes

Below are 10 best foods to prevent and control diabetes:

1. Bitter gourd (bitter melon)

Bitter gourd (bitter melon) contains compounds that help lower blood glucose and fats levels in the body. Bitter gourd juice is an excellent beverage for people with diabetes. In fact, it is more effective than some second-line drugs for controlling glucose levels in the body. It works through various mechanisms to lower blood sugar levels. It limits the breakdown of the carbohydrates into glucose by inhibiting the carbohydrates metabolizing enzymes.

Furthermore, bitter gourd enhances the glucose uptake by tissues and increases glucose metabolism. It repairs damaged beta cells that make insulin and prevents their death. It contains chemical compounds such as charantin and polypeptide-p that exhibit a hypoglycemic effect. Polypeptide-p or p-insulin is an insulin-like protein. It works by mimicking the action of insulin in the body and very effective in controlling sugar levels in patients with type-1 diabetes.

Bitter gourd helps in treating obesity by boosting the system and enzymes responsible for converting fat into energy. It prevents the accumulation of fat in the body, which prevents fat-induced insulin resistance. In season, you must eat at least one medium bitter gourd or 50-100 ml of bitter gourd juice in a day. Bitter gourd could be your magic pill to reduce your dependence on your anti-diabetic medications. If you are healthy and young but have a family history of diabetes, then you should start eating bitter gourd to prevent diabetes in the future.

2. Fenugreek seeds

Fenugreek seeds are the second most effective food, after bitter gourd to control diabetes naturally. Regular consumption of fenugreek seeds effectively prevents the development of diabetes. Fenugreek seeds increase glucose-induced insulin release. The research finding shows that after consumption of fenugreek seeds soaked in hot water, significantly decrease fasting blood sugar, triglyceride, and LDL-cholesterol up to 30%. If you have diabetes, you should eat fenugreek seeds every day. But before you start eating them, consult your doctor because regular consumption of fenugreek seed decreases your blood sugar levels, and you need a lesser dose of your prescribed drug. Soak fenugreek seeds in a cup of water overnight. The next morning on an empty stomach, chew the seeds and drink the water in which the seeds were soaked.

3. Bottle gourd/Calabash

Consumption of bottle gourd helps reduce blood sugar levels. Bottle gourd is very low in calories and high in both soluble and insoluble dietary fiber. It contains almost 90% water, which makes it a choice of vegetable in diabetes. Bottle gourd helps prevent the development of insulin resistance in type 2 diabetes. It inhibits the action of an enzyme called protein-tyrosine phosphatase (PTP) 1B, which improves glucose metabolism and enhances insulin sensitivity without causing lipid accumulation in the liver and thus helps in obesity control.

Make sure you don't eat bitter bottle gourd. First taste a piece of bottle gourd before cooking, discard it if it is bitter because bitter bottle gourd is not edible and can even cause toxicity and stomach ulcer.

4. Barley

If you want to prevent diabetes, start eating barley regularly. The lower consumption of dietary fiber is associated with the increasing prevalence of diabetes. Barley is an excellent source of soluble fiber along with antioxidant minerals such as magnesium, copper, selenium, and chromium. Research shows long term consumption of barley is effective in lowering blood glucose levels by mimicking the mechanism of action of your first line anti-diabetes drugs. It decreases insulin resistance and interferes with carbohydrates absorption and metabolism. Carbohydrates in barley convert into glucose gradually, without rapidly increasing blood glucose levels. It increases a hormone that helps reduce chronic low-grade inflammation. Barley is an incredible preventive food for those who are at high risk for developing diabetes. You can grind barley and make flour. Add barley flour in wheat flour whenever you make chapati or bread; you can even add barley flour to your cake batter.

5. Monounsaturated fats

Monounsaturated fats such as olive oil, canola oil, and avocado can be advantageous for those with type 1 or type 2 diabetes who are trying to lose or maintain weight. High-monounsaturated-fat diets cause a modest increase in

HDL-cholesterol levels, and lower LDL-cholesterol levels, as well as improve glycemic control. Oil containing monounsaturated fats is the choice of oil in diabetes, so choose your cooking oil accordingly. You can protect your heart by replacing saturated fats in your diet with monounsaturated fats. Regular consumption of monounsaturated fats prevents insulin resistance and accumulation of abdominal fat by increasing your fat-burning hormone adiponectin. Keep in mind that oils are high in calories, even the healthiest oils, so consume them in moderation. Your objective should be to replace saturated fats with monounsaturated fats. Keep the intake of polyunsaturated fats (soybean oil, sunflower oil, and corn oil) less than 10% of total energy consumption. Your total fat consumption should be less than 35% of total energy consumption (from carbohydrates and protein).

6. Legumes

Legumes are a superfood for people with diabetes. Legumes such as chickpeas, kidney beans, and peas help manage and reduce type 2 diabetes risk. They are low on the glycemic index (GI) scale despite containing carbohydrates. They increase serum adiponectin concentrations in type 2 diabetic patients that help in preventing abdominal fat and reduce the risk of insulin resistance. Always choose dried beans over canned beans because lots of salt is added in canned products, which can increase your risk of high blood pressure. If you must use canned beans, be sure to rinse them to get rid of salt as much as possible.

7. Zinc

Zinc plays an antioxidant role in type 2 diabetes. It improves oxidative stress by reducing chronic hyperglycemia. It has been seen that people with diabetes have lower levels of zinc than people without diabetes. Zinc deficiency may lead to the development of diabetes. It is because zinc plays a crucial part in insulin metabolism; it helps in the production and secretion of insulin. As zinc strengthens the immune system, it protects beta cells from destruction.

Moreover, zinc prevents diabetes by increasing the levels of adiponectin hormone in the body that helps reduce weight. Studies suggest zinc-rich foods help lower blood sugar levels in type 1 as well as type 2 diabetes. Foods that are high in zinc are cashew nuts, sesame seeds, chickpeas, kidney beans, milk, and oats.

8. Fruits

Because of the sugar content, generally, people with diabetes avoid fruits, which is not right. Fruits are full of soluble fiber and don't contain free sugar that is found in chocolate, cakes, biscuits, fruit juices, and cold drinks. So, if you want to cut down the sugar intake, avoid fruit juices, sugary drinks, and cakes than whole fruits. You can easily have one large banana or a medium apple or one slice of papaya in a day.

9. Low-fat yogurt

Probiotics help to reduce inflammation in the body. Yogurt is the best example of probiotics. Yogurt is low in carbohydrates and contains a good amount of protein, vitamin D, calcium, and potassium. It reduces fasting blood glucose, blood pressure, lipid profile, and other cardiovascular risk factors in people with type 2 diabetes. Probiotics control the glycemic condition by lowering insulin resistance and decreasing the production of inflammatory markers. People who eat yogurt have better control over blood sugar in comparison to those who do not eat yogurt. Choose low-fat yogurt over regular yogurt to prevent weight gain.

10. Indian gooseberry/ Amla

The Indian gooseberry is the richest source of vitamin C, containing 20 times more vitamin C than that of orange. Vitamin C lowers blood pressure in people with type 2 diabetes and protects your heart. Indian gooseberry is rich in nutrients and phytochemicals, such as gallic acid, ellagic acid, gallotannin, and corilagin. These all are potent antioxidants. Through their free radical scavenging properties, these phytochemicals help prevent and control hyperglycemia, cardiac complications, and diabetic complications like nephropathy and neuropathy. Amla favorably impacts on the lipid profile, it significantly elevates HDL-cholesterol and lowers LDL-cholesterol levels. Eat it raw or make chutney or simply boil and consume the strained amla juice.

CONCLUSION

Type 2 diabetes is a lifestyle disease. The best part is you can prevent and control diabetes by making some lifestyle modifications. Lack of awareness and over-promotion of unhealthy foods and bad lifestyle choices are the reason why the incidence of diabetes is increasing in population. There is no harm in enjoying unhealthy foods, considering you are consuming them in moderation. Whether or not your blood sugar level is high, to effectively prevent diabetes, reduce your frequency of eating out. Eat homemade foods. Make every unhealthy food at home, and make it from scratch. By the end, when your food will be ready to eat, you will involuntarily fill with the guilt of eating so many unhealthy things at a time. From the next time, you will crave less. Try this trick, it works!

If you are on diabetes medications, consult your doctor and pharmacist before adding the foods mentioned above in your diet. They can provide you the best advice about which foods you can eat and which ones not, given your diabetes condition and the complications of your diabetes. When you eat the foods mentioned above, your blood glucose levels drop, and you need a lesser dose of medicines. So, regularly discuss with your doctor to adjust the dose accordingly, sometimes doctors just repeat the previous prescription in a hurry without checking your current blood sugar levels.

EAT TO PREVENT AND CONTROL DIABETES

KEY POINTS

- ✓ Regularly check your blood pressure. Maintain blood pressure <130/80 mm Hg. Tight control of blood pressure can be more effective than glycemic control in preventing heart disease.

- ✓ Be active. Play outdoor games, use stairs, don't sit for long. A sedentary lifestyle leads to obesity, which causes insulin resistance.

- ✓ Increase adiponectin (fat burning hormone) levels in your body by eating monounsaturated fats.

- ✓ Don't just add additional monounsaturated fats in your diet, instead replace saturated fats with monounsaturated fats.

- ✓ Drink plenty of water. Water helps remove excess sugar from your blood through urine.

- ✓ Eat foods that contain soluble fiber and complex carbohydrates that are low in the glycemic index. You can typically eat protein, but proteins are restricted in people who have or are at risk of kidney damage.

- ✓ Keep an eye on your hypoglycemia symptoms. Always keep some glucose tablets with you.

- ✓ If you have diabetes, avoid diabetes complications by getting regular diabetes tests, which can catch problems early and can help you prevent serious diabetes complications. Get a yearly eye exam to ensure no blood vessels of the retina has damaged. Get your cholesterol levels checked. Get a regular urine microalbumin tests to check your kidneys' health, and, electrocardiogram to check your heart health.

UNIT 3

DIET PLAN

Diet Plan for

Diabetes

Diabetes + Hypertension

Diabetes + Arthtitis

DIET PLAN TO CONTROL DIABETES

Diet plan to control diabetes

- Soak one tablespoon of fenugreek seeds overnight in 250 ml of water. The next morning chew these seeds and drink the fenugreek water. Do this every day (highly effective).

EAT TO PREVENT AND CONTROL DIABETES

- Chew some (3-4) basil leaves or add them in your morning green tea.

- Eat a handful of overnight soaked nuts.

- Replace potatoes with sweet potatoes and white rice with brown rice.

- Eat one crushed garlic on an empty stomach thrice a week. (after one hour of having soaked fenugreek seeds).

- Add barley flour to your whole wheat flour in a ratio of 1:7. Add 1 kg barley flour in 7 kg of whole wheat flour. Beta-glucan of barley is very effective in preventing diabetes and also prevent weight gain.

- Drink more water, about 2-3 liters, equivalent to 10 -12 glass of 250 ml. Water helps remove excess sugar from your blood through urine, and it helps prevent dehydration.

- In season, drink 50 ml to 100 ml of fresh bitter gourd juice every day. Cook bitter gourd with its peel. Don't remove the skin.

- Eat a variety of sprouts every day.

- Add flax seeds in your dough or add them in yogurt.

- Eat vegetables that have high water content such as bottle gourd and ridge gourd.

- Eat non-starchy vegetables such as carrot, cabbage, cauliflower, green beans, and broccoli.

- Eat vitamin C foods such as amla, lemon, orange, and capsicum.

- Use olive oil, canola oil, and mustard oil in cooking.

- Include apple, oatmeal, and beans in your diet for high soluble fiber.

DIET PLAN TO CONTROL DIABETES + HYPERTENSION

Diet plan to control diabetes + hypertension

- Drink warm lemon water on an empty stomach.

- After half an hour, eat soaked fenugreek seeds and drink the fenugreek water. Do this every day.

- After one hour, eat one crushed garlic. Do this every day.

- Have green tea with added lemon and basil leaves.

- Eat a handful of overnight soaked nuts.

- Drink about 2-3 liters of water in a day.

- In season, drink 50 ml to 100 ml of fresh bitter gourd juice every day.

- Add barley flour to your whole wheat flour in a ratio of 2:10. Add 200 gm of barley flour to 1 kg of whole wheat flour.

- Eat one banana, especially if you are taking hypertension medicines. Don't remove the banana from your diet just because you have diabetes. Cut the sugar intake in tea and other high glycemic fruits.

- In season, eat multigrain beetroot paratha (see the recipes section). Have 50 to 100 ml of beetroot juice every day.

- Eat sprouts every day.

- Add flax seeds in dough or add them in a yogurt fruit salad.

- Eat bottle gourd, carrot, and ridge gourd.

- Replace potatoes with sweet potatoes and eat them in moderation.

- Drink low-fat cow's milk boiled with turmeric powder at night.

- Eat plenty of spinach, kale, cabbage, and chenopodium (bathua).

- Increase legumes consumption, including lentils, chickpeas, kidney beans, and soybeans.

- Eat vitamin C foods such as amla, lemon, orange, and capsicum.

- Use olive oil, canola oil, and mustard oil in cooking.

DIET PLAN TO CONTROL DIABETES + ARTHRITIS

Diet plan to control diabetes + arthritis

- Soak one tablespoon of fenugreek seeds overnight in 250 ml of water. The next morning chew these seeds and drink the fenugreek water. Do this every day. (highly effective)

- Eat a handful of overnight soaked walnuts and dried figs (2 pieces). Eat them daily.

- Drink green tea mixed with crushed ginger and basil leaves (3-4 leaves).

- Eat one crushed garlic on an empty stomach thrice a week (after one hour of having soaked fenugreek seeds).

- Add barley flour and soybean flour to your whole wheat flour in a ratio of 1.5:1:10. Add 1.5 kg barley flour and 1 kg of soybean flour in 10 kg of whole wheat flour.

- Replace potatoes with sweet potatoes and white rice with brown rice.

- Drink 2-3 liters of water in a day.

- Add horse gram (kulthi) in your diet.

- In winter, eat plenty of fresh turmeric roots. Drink low-fat cow's milk boiled with turmeric powder at night.

- In season, drink 50 ml to 100 ml of fresh bitter gourd juice every day. Cook bitter gourd with its peel. Don't remove the skin.

- Eat sprouted horse gram, sprouted mung bean, and black chickpeas.

- Add flax seeds in dough or add them in the yogurt fruit salad.

- Eat vegetables that have high water content such as bottle gourd and ridge gourd.

- Eat non-starchy vegetables such as carrot, cabbage, cauliflower, green beans, and broccoli.

- Eat plenty of green vegetables including, spinach, kale, and fenugreek leaves.

EAT TO PREVENT AND CONTROL DIABETES

- Use olive oil, canola oil, and mustard oil in cooking, and avoid sunflower oil and corn oil.

- Include apple, oatmeal, and beans in your diet for high soluble fiber.

UNIT 4

RECIPES

Healthy and tasty recipes to boost your health

LUNCH

CHOLE MASALA

NON-FRIED OATS BHATURE

CHOLE MASALA

Chole masala

Serves 4

Ingredients:

To cook chickpeas

Uncooked dried Chickpeas: 200 grams

Teabags: 3

Onion: 1 medium

Salt: To taste

Water: 500 ml

For gravy

Garlic: 12 cloves

Ginger: 2 inches

Onion: 3 medium size

Asafoetida: ½ teaspoon

Cumin: 1 teaspoon

Bay leaves – 1

Coriander Powder- 1 teaspoon

Turmeric powder: 1 teaspoon

Red chili powder: 1 teaspoon

Garam Masala- 1 teaspoon

Chole Masala- 2 tablespoons

Dry mango powder: 1 teaspoon

Salt: To taste

Water: 300 ml

Mustard oil: 2 tablespoons

Method:

1. Soak the chickpeas in enough water. Keep it overnight or for at least 8 hours.

2. Next morning, drain the water and rinse chickpeas thoroughly.

3. In a pressure cooker, add the soaked chickpeas, tea bags, salt, sliced onion (1 medium), and 500 ml water. Pressure cook at medium flame for 5-6 whistles. It will take around 15 minutes. The chickpeas should be soft when you mash it with a spoon.

4. If chickpeas are hard, pressure cook them for two whistles.

5. Strain the chickpeas and keep the stock for later use.

6. Grind onion, garlic, ginger, together to make a smooth paste.

7. Add mustard oil to the hot pan, add asafoetida, cumin, and bay leaves. Sauté for 2-3 minutes.

8. Add the onion mixture in the oil. Mix well and cover the pan with a lid. Cook at the medium-low flame for 15 minutes. Cook till it leaves oil. The raw taste of onion should evaporate completely.

9. Add turmeric powder, coriander powder, dry mango powder, red chili powder, salt (keep in mind, we have added salt while cooking chickpeas too), garam masala, and chole masala (skip it if you don't have chole masala). Add another teaspoon of garam masala if you are not adding chole masala.

10. Cook for 2-3 minutes.

EAT TO PREVENT AND CONTROL DIABETES

11. Add chickpeas, mix well. Masala mixture should coat the chickpeas entirely.

12. Cover it and cook for 5-7 minutes.

13. Add the stock in which chickpeas were cooked. Add another 300 ml of water. Water should be just above the chickpeas. Don't add too much water, it will dilute the taste.

14. Take a masher and mash almost 20% of chickpeas to make the gravy thick.

15. Cover it and cook for 18-20 minutes on low flame until chickpeas absorb the flavor of masala. The gravy should be thick.

16. Your chole masala is ready to eat. Eat it with non-fried oats bhature or brown rice.

NON-FRIED OATS BHATURE

Non-fried oats bhature

Serves 4

Ingredients:

Oats: 2 ½ cup

Whole wheat flour: 1 ½ cup

Salt: To taste

Baking powder: 1 teaspoon

Baking soda: 1 teaspoon

Thick yogurt: 1 cup

Olive oil – 1 tablespoon + for making bhature

Method:

1. Grind the oats. Mix with whole wheat flour in a large bowl. Add salt, baking powder, baking soda, and oil. Mix well.

2. Add 1 cup of thick yogurt and knead for 5-6 minutes. The dough should be a little stiff for making crispy bhature. If required, add more yogurt.

3. Cover the dough with a wet muslin cloth. Leave it for at least 2 hours.

4. Divide the dough into 15 equal parts.

5. Grease your palm well with oil. Take 1 part and make a ball shape between your palm. Smoothen the ball, make sure it's crack free.

6. Repeat the process for all parts.

7. Now take 1 ball and roll it into oval shape or round discs. It should neither be too thick nor thin.

8. Heat a griddle. Grease it with oil. Put bhatura on a hot greased griddle.

9. First, cook one side. Turn the bhatura and add 1 teaspoon of oil to make it crisp from the other side too. Cook the other side. Brown spots should appear on both sides.

10. Repeat with the remaining dough to make 14 more bhature.

11. Eat non-fried oats bhature with chole masala.

NOTE FROM LA FONCEUR

Dear Reader,

Thank you for reading *Eat to Prevent and Control Diabetes*. I hope you have found this book helpful.

If you have a moment, please leave a review online. Help other health-conscious readers find this book by telling them why you enjoyed reading. You can review *Eat to Prevent and Control Diabetes*.

Join my mailing list at www.eatsowhat.com/mailing-list

Learn how a vegetarian diet is the solution to a disease-free healthy life in *Eat So What!* series- *Eat So What! The Power of Vegetarianism* and *Eat So What! Smart Ways to Stay Healthy*.

If you are looking for a permanent solution to your hair problems, read my book *Secret of Healthy Hair*.

All of my books are available in eBook, paperback, and hardcover editions.

Regards

La Fonceur

REFERENCES

1. Soheil Z, Habsah A, "A Review on Antibacterial, Antiviral, and Antifungal Activity of Curcumin." Biomed Res Int. 2014; 2014: 186864.
2. Silagy C, Neil A, "Garlic as a lipid lowering agent--a meta-analysis." R Coll Physicians Lond. Jan-Feb 1994;28(1):39-45.
3. Matthias B, Mandy S, "Fiber and magnesium intake and incidence of type 2 diabetes: a prospective study and meta-analysis." Arch Intern Med. 2007 May 14;167(9):956-65.
4. Karin R, Toben C, Fakler P, "Effect of garlic on serum lipids: an updated meta-analysis." Nutr Rev. 2013 May;71(5):282-99.
5. Holly L, Sharon A, "Garlic and onions: Their cancer prevention properties." Cancer Prev Res (Phila). 2015 Mar; 8(3): 181-189.
6. Ranade M, Mudgalkar N, "A simple dietary addition of fenugreek seed leads to the reduction in blood glucose levels: A parallel-group, randomized single-blind trial." Ayu. 2017 Jan-Jun; 38(1-2): 24-27.
7. Calado A, Neves M, "The Effect of Flaxseed in Breast Cancer: A Literature Review." Front Nutr. 2018; 5: 4.
8. Chikako M, Taeko K, "Effects of glycolipids from spinach on mammalian DNA polymerases." Biochem Pharmacol. 2003 Jan 15;65(2):259-67.
9. Mondal S, Varma S, "Double-blinded randomized controlled trial for immunomodulatory effects of Tulsi (Ocimum sanctum Linn.) leaf extract on healthy volunteers." Ethnopharmacol. 2011 Jul 14;136(3):452-
10. Dokania M, Kishore K, Sharma PK, "Effect of Ocimum sanctum extract on sodium nitrite-induced experimental amnesia in mice." Thai J Pharma Sci. 2011; 35:123-30.
11. Eddouks M, Amina B, "Antidiabetic plants improving insulin sensitivity." J Pharm Pharmacol. 2014 Sep;66(9):1197-214.
12. Widjaja S, Rusdiana, "Glucose Lowering Effect of Basil Leaves in Diabetic Rats." J Med Sci. 2019 May 15; 7(9): 1415-1417.
13. A review on, "What You Need to Know about Dietary Supplements." https://www.fda.gov/food/
14. Sarwar N, Gao P, "Diabetes mellitus, fasting blood glucose concentration, and risk of vascular disease: a collaborative meta-analysis of 102 prospective studies. Emerging Risk Factors Collaboration." Lancet. 2010; 26; 375:2215-2222.
15. Bourne R, Stevens G, "Causes of vision loss worldwide, 1990-2010: a systematic analysis." Lancet Global Health 2013;1:e339-e349

16. "Diabetes facts & figures." The International Diabetes Federation (IDF), 2020, https://www.idf.org/aboutdiabetes/what-is-diabetes/facts-figures.html

17. Cruz K, Oliveira A, "Antioxidant role of zinc in diabetes mellitus." World J Diabetes. 2015 Mar 15; 6(2): 333-337.

18. Abdelsalam S, Hesham M, "The Role of Protein Tyrosine Phosphatase (PTP)-1B in Cardiovascular Disease and Its Interplay with Insulin Resistance." Biomolecules. 2019 Jul; 9(7): 286.

19. Alam MD, Uddin R, "Beneficial Role of Bitter Melon Supplementation in Obesity and Related Complications in Metabolic Syndrome." J Lipids. 2015; 2015: 496169.

20. Padmaja Chaturvedi P, "Antidiabetic potentials of Momordica charantia: multiple mechanisms behind the effects." J Med Food. 2012 Feb;15(2):101-7.

21. Joseph B, Jini D, "Antidiabetic effects of Momordica charantia (bitter melon) and its medicinal potency." Asian Pac J Trop Dis. 2013 Apr; 3(2): 93-102.

22. Tayyab F, Lal S, "Medicinal plants and its impact on diabetes." World J Pharm Res. 2012;1(4):1019-1046.

23. Paul A, Raychaudhuri S, "Medicinal uses and molecular identification of two Momordica charantia varieties - a review." E J Bio. 2010;6(2):43-51.

24. Garg A, "High-monounsaturated-fat diets for patients with diabetes mellitus: a meta-analysis." Am J Clin Nutr, 1998 Mar;67(3 Suppl):577S-582S.

25. Kassaian N, Azadbakht L, "Effect of fenugreek seeds on blood glucose and lipid profiles in type 2 diabetic patients." Int J Vitam Nutr Res. 2009 Jan;79(1):34-9.

26. Kesika P, Sivamaruthi B, "Do Probiotics Improve the Health Status of Individuals with Diabetes Mellitus? A Review on Outcomes of Clinical Trials." Biomed Res Int. 2019; 2019: 1531567.

27. D'souza J, D'souza P, "Anti-diabetic effects of the Indian indigenous fruit Emblica Officinalis Gaertn: active constituents and modes of action." Food Funct. 2014 Apr;5(4):635-44.

28. Sharabi K, Tavares C, "Molecular Pathophysiology of Hepatic Glucose Production." Mol Aspects Med. 2015 Dec; 46: 21-33.

29. Santhi K, Rajamani S, "Amla, a Marvelous Fruit for Type -2 Diabetics-A Review." Int. J. Curr. Microbiol. App. Sci. 2017, 5:116-123

30. Radzeviciene L, Ostrauskas R, "Adding Salt to Meals as a Risk Factor of Type 2 Diabetes Mellitus: A Case-Control Study." Nutrients. 2017 Jan; 9(1): 67.

31. Li J, Zhang N, "Non-steroidal anti-inflammatory drugs increase insulin release from beta cells by inhibiting ATP-sensitive potassium channels." Br J Pharmacol. 2007 Jun; 151(4): 483-493.

IMPORTANT TERMINOLOGY

Arteries: Arteries are blood vessels that carry oxygen-rich blood from the heart to the body.

Vascular: Vascular is related to vessels that carry blood in the body.

Glucose: Glucose means sweet in Greek. It is a type of sugar. You get carbohydrates from foods, which break down into glucose, and your body uses it for energy.

Hyperglycemia: Hyperglycemia refers to high levels of sugar (glucose) in the blood.

Bioavailability: The actual proportion of a substance that reaches the blood circulation after it introduced into the body to show its effect.

Free radicals: Free radicals are the unpaired electrons that form due to the oxidative process in the body. These unpaired electrons like to be in pairs, so they pair with the electrons in proteins and DNA and damage them.

Antioxidants: Body has antioxidants, which neutralize free radicals by inhibiting the oxidative process that forms free radicals.

Oxidative stress: When free radicals outnumber the naturally occurring antioxidants, it results in oxidative stress. This imbalance leads to cell and tissue damage, including DNA, protein, and lipids. Damage to your DNA increases your risk of chronic diseases such as cancer, rheumatoid arthritis, diabetics, stroke, and aging.

Inflammation: Inflammation is the response of the body to harmful pathogens, and irritants, and eliminate the initial cause of cell injury. Body releases white blood cells to heal the damaged cells. When the immune system mistakenly attacks healthy tissue, it causes harmful abnormal inflammation. Reasons for abnormal inflammation are stress, smoking, and alcohol consumption. Some examples of diseases associated with abnormal inflammations are rheumatoid arthritis, psoriasis, and inflammatory bowel diseases.

ABBREVIATIONS

RA- Rheumatoid Arthritis

BP- Blood pressure

LDL- Low-density lipoproteins

HDL- High-density lipoprotein

COX- Cyclooxygenase

PG- Prostaglandin

RAAS- Renin–angiotensin–aldosterone system

ABOUT THE AUTHOR

La Fonceur is the author of the book series *Eat So What!* and *Secret of Healthy Hair*, a dance artist, and a health blogger. She has a master's degree in Pharmacy, and she specialized in Pharmaceutical Technology. She has published an article titled 'Techniques for Producing Biotechnology-Derived Products of Pharmaceutical Use' in Pharmtechmedica Journal. She is a registered state pharmacist. She is a national-level GPAT qualifier of the year 2011 in which she was among the top 1400 nationwide. Being a research scientist, she has worked closely with drugs. Based on her experience, she believes vegetarian foods are the remedy for many diseases; one can prevent most of the diseases with nutritional foods and a healthy lifestyle.

ALL BOOKS BY LA FONCEUR

Full-length books:

Mini editions:

Hindi editions:

CONNECT WITH LA FONCEUR

Instagram: @la_fonceur | @eatsowhat

Facebook: LaFonceur | eatsowhat

Twitter: @la_fonceur

Amazon Author Page:

www.amazon.com/La-Fonceur/e/B07PM8SBSG/

Bookbub Author Page:
www.bookbub.com/authors/la-fonceur

Sign up to my website to get exclusive offers on my books:

Blog: www.eatsowhat.com

Website: www.lafonceur.com/sign-up

Lightning Source UK Ltd.
Milton Keynes UK
UKHW021200190121
377256UK00001B/27